Liset Barreda Jorge
Dayamí Fontes Marrero
Lissette Arzola de la Rosa

FOOD SOVEREIGNTY OF THE MUNICIPALITY OF MORÓN

Liset Barreda Jorge
Dayamí Fontes Marrero
Lissette Arzola de la Rosa

FOOD SOVEREIGNTY OF THE MUNICIPALITY OF MORÓN

A PROPOSAL TO STRENGTHEN THE PRINCIPLES OF THE LOCAL AGRICULTURAL INNOVATION SYSTEM IN CIEGO DE ÁVILA. CUBA

ScienciaScripts

Imprint

Cover image: www.ingimage.com

This book is a translation from the original published under ISBN 978-3-639-53669-0.

Publisher:
Sciencia Scripts
is a trademark of
Dodo Books Indian Ocean Ltd. and OmniScriptum S.R.L publishing group

120 High Road, East Finchley, London, N2 9ED, United Kingdom
Str. Armeneasca 28/1, office 1, Chisinau MD-2012, Republic of Moldova, Europe
Printed at: see last page
ISBN: 978-620-5-93926-0

A PROPOSAL FOR STRENGTHENING THE PRINCIPLES OF SIAL IN FOOD SOVEREIGNTY IN THE MUNICIPALITY OF MORÓN.

AUTHORS: DR. C. LISET BARREDA JORGE DR. C. DAYAMÍ FONTES MARRERO
DR. C. LISSETTE ARZOLA DE LA ROSA

SUMMARY

In the agricultural context of the municipality of Morón there are several local enterprises that constitute innovations that typify the locality, however, the articulation of the agents of change in the agricultural context is insufficient, there is a lack of perception among the actors about local development and the awareness of the actors about the need to form innovative capacities in producers, academics, scientists and decision makers to continue at a more strengthened level is deficient, which is why the scientific problem of how to contribute to Food Sovereignty in the municipality of Morón was determined, The objective was to design a multi-stakeholder action plan based on the strengthening of the principles of SIAL to contribute to Food Sovereignty in the municipality of Morón. The application of various instruments and consultation with experts made it possible to identify the strengths, weaknesses, threats, and opportunities of the agricultural and innovation context, leading to the construction of their respective SWOT matrices, which show their influence on the innovation systems. A multi-stakeholder action plan was then drawn up that stands out for being staggered, integrated, contextualized and articulating the mechanisms that guarantee the necessary coherence between local stakeholders such as producers, academics, scientists and decision-makers, with the intention of strengthening the principles of the SIAL, and thus contributing to Food Sovereignty in Morón, from the perspective of social equity.

INDEX

INTRODUCTION

To face the effects of the world crisis, in Cuba, economic actions are developed in search of rational solutions in favor of improving the quality of life of the population, that is why, to sensitize the society is a necessity of these times and means to preserve the resources and identity values, with the active participation of all the actors, looking for new solutions and alternatives in the local space and thus contribute to improve the coexistence.In the last 20 years there has been a vertiginous development in studies with an impact on agricultural production, since this sector can contribute to the achievement of social, environmental and economic objectives. Socially, by producing nutritious and safe food and reducing health risks. Environmentally, by making efficient use of renewable and non-renewable resources, reducing soil erosion, maintaining or improving soil quality, and minimizing the risk of water pollution. Economically, by generating wealth, food trade and jobs that can benefit socially disadvantaged groups. In view of the above, it can be said that achieving agricultural development in a municipality is one of the main tasks that local governments must address in order to strengthen local development. Rodríguez, A. (2020)On the other hand, the economic situation invites us to reflect on the perspectives and importance of local development, even more so, when the global economic crisis together with the financial, political and environmental crises, symbolize the end of a growth cycle and the moment to synthesize on the results of the policies that have led and driven it by seeking alternative events to safeguard the human species and achieve a fairer, more equitable and sustainable world.Cuba is immersed in a process of economic decentralization of implementation of an Economic Model in which priority is given to local municipal governments as active protagonists, managers of their own development, which brings with it the level of management and articulation between the agents of change to provide important endogenous transformations at the local level. Guidelines that deal with municipal issues in this sense are: 21, 36, 37, 146, 155 and 164. Communist Party of Cuba, (2011)For a municipality to develop integrally, it must take full advantage of the internal and external potentialities of the territorial context in which it develops, which is one of the fundamental premises of the guidelines of the country's Economic Policy, approved by the VI Congress of the Communist Party of Cuba (PCC).This use cannot be seen as alien to the proposed objectives, if the municipality does not have the human resources to achieve the necessary and sufficient transformations at the local level, but above all to work towards achieving Food Security or Food Sovereignty, in accordance with the needs of the municipality.Cuba's Food Sovereignty and Nutrition Education Plan constitutes the national platform to achieve full Food Security; it was carried out through participatory activities with a gender and generational approach, and considering the Economic and Social Policy Guidelines of the Party and the Revolution for the period 2016-2021, the Bases of the National Economic and Social Development Plan until 2030 and the Sustainable Development Goals. Plan for Food Sovereignty and Nutritional Education in Cuba (2020).Morón is a municipality located in the north and center of the province of Ciego de Avila, between 780 17' and 780 54' west longitude and 22o 00' and 220 40' north latitude; it has a territorial extension of 598.80 km2 on land and 652.82 km2 of keys (Coco, Guillermo, Paredón Grande, Antón Chico and Romano), which makes a total area of 1 251.62 km2. It is bordered on the east by the municipality of Bolivia, on the north by the Old Bahamas Channel, on the west by the municipalities of Chambas and Ciro Redondo, and on the south by the municipalities of Ciro Redondo and Primero de Enero (Figure 1). (Figure 1).The municipality is composed of 6 Popular Councils, 7 Defense Zones and 72 districts with a population of 70,595 inhabitants with 2 urban and 17 rural settlements, the economy of the territory is composed by the activity of services; tourism, health and education with several facilities of regional scope.In agriculture for various crops, pasture, forestry, as well as fruit and sugar cane cultivation. In industry; construction materials, fruit and vegetable canning, machinery construction, textile manufacturing, food industry, light industry, aquaculture and fishing. In addition, there are two scientific centers that contribute to the following science and technology transfer in conjunction with a Municipal University Center.The Júcaro-Morón Plain occupies most of the territory, made up of flat, gently undulating and hilly plains; towards the coasts the plains are low, accumulative swamps with mangrove vegetation. Both coasts are bordered by an arc of islands and keys that form shallow seas and the different types of soils that are presented are linked to the topography, being of the type ferralíticas Red, deep of good drainage, so much internal as external and little risk of erosion.

Figure 1. Location and boundaries. Morón Municipality

Source: Municipal Statistics Office (OMEI) of Morón.

It is located in the Júcaro to Morón Plain, where flat plains predominate (0-33 m), corresponding to 84% of the territory, while 6% belongs to small elevations (Turiguanó Island, maximum peak 105.3 masl), and 10% are swampy areas that make up the Great Northern Wetland.From a productive point of view, the agricultural area is used for permanent crops (3,905.25 ha) (mainly sugar cane and fruit trees), and for temporary crops (2,906.49 ha) (mainly grains and vegetables).In Morón there are antecedents of intentionality for the implementation of the Local Agricultural Innovation System (SIAL) as a result of the research carried out as part of the first and second edition of the diploma courses in Ciego de Avila Province, as Barreda. L, Bilbao. B, Martínez. Y, Castellanos. L, González. K. (2018) and Ramírez. B, Duharte. S, Mundiel. D, Jiménez. C and Arzola. L, (2019), in 2018 the former proposed a multi-stakeholder action plan for inserting the System of Local Agricultural Innovation in the municipality and in 2019 the second ones designed another action plan for the creation of the Local Agricultural Innovation Group (GIAL), in the productive bases of the municipality, by taking into account the production systems that integrate agriculture-livestock farming. The SIAL is a mechanism that enables local government and the local stakeholders involved to resolve the barriers that hinder the development of agrifood chains at the local level. This system is composed of the Multi-stakeholder Management Platform (PMG) and the Local Agricultural Innovation Groups (GIAL) that function by articulating diverse local actors and local government (Ortiz, R, la O, M and Miranda; 2017).

It is significant to note, in the current context, that the advance of the coronavirus in Morón, has forced to isolate communities, and to protect vulnerable people, such as the elderly and patients with chronic diseases, poses an added challenge for food production, which it has been trying to ensure since 2016 through the Municipal Self-Supply Program, as part of the country's sustainable development goals. Tamayo. A, (2020)However, this has not stopped the intentionality of the implementation of the SIAL in Morón and, during a check-up of the municipal self-supply program, with the participation of sector leaders and producers, Valdés Mesa, highlighted the growth of more than two thousand hectares of crops in the territory and the existing conditions to consolidate agricultural production despite weather setbacks and shortages of inputs. Zamora. M, (2021During the same visit, Valdés Mesa urged the cooperatives dedicated to sugarcane not to limit themselves to this crop and to seek new income and sustainability in other variants such as the planting of viands and animal husbandry, while he called to extend the experience of several productive bases to close the cycle from the harvest to the production of new products in the mini-industries and the linking of sales through the export pole. Zamora. M, (2021The Local Agricultural Innovation System (SIAL) is part of a project led by a group of researchers from the National Institute of Agricultural Sciences (INCA) and financed by a Swiss Cooperation for Development in Cuba.The implementation of the innovation promotion paradigm in its very essence is an innovative act that requires learning by doing and demands the participation of all local actors, accompanied by the academy.The SIAL should be steered by the municipal government, which can also take on planning and systematic checking roles, as well as encouraging the participation of all the actors that make important decisions in the municipality and thus mitigate the obstacles that have arisen up to now and that can be resolved at the local level.

The SIAL is a proposal on "how" to implement a knowledge management and development system in the municipalities, with attributes of horizontality and participation.As can be seen in Figure 2, the system promotes forms of social organization of innovation such as: Multi-stakeholder Management Platforms (PMG) and Local Agricultural Innovation Groups (GIAL).These components give rise to cycles of knowledge management and innovation based on learning by doing. In the operation of the SIAL, capacities are strengthened to legitimize the principles and good practices of local and participatory innovation.Local agro-food development, in this case in the municipality of Morón, requires the formation of individual and collective subjects appropriate to a conception of sustainable development on agroecological bases and a gender equity approach, which makes it possible to generate interactive learning processes through an intentional dialogue between scientific knowledge and peasant knowledge.This implies the design of dialogic, participatory educational processes that include theoretical conceptions and methodologies that favor the exchange of experiences.Training these skills requires a program that provides participants with pedagogical conceptions that privilege the collective construction of knowledge, horizontality in the educator/ra-educatee relationship, and the conformation of the group as the subject of the learning process.The lessons learned, as a cycle of learning, in the Local Agricultural Innovation System (SIAL) have facilitated the apprehension of a proactive ideology that is specifically translated into the requirements of social innovation for the contribution, concretion and implementation of good practices that have an impact on food security in the municipality.

Figure 2. Local agricultural innovation system.

Source: Taken from texts supporting the Diploma course for the implementation of the SIAL.

The experience in the municipality of Morón, which recently joined the SIAL, although it is not one of the municipalities prioritized in the province within the national project, highlights the need to continue strengthening the innovative capacities of producers, academics, scientists and decision-makers to reaffirm the implementation of the SIAL with objectivity and sustainability.This municipality requires the coordination of stakeholders and decision-makers to strengthen the SIAL and thus achieve coordination, planning, implementation, follow-up, and evaluation of agricultural research and innovation processes in an integrated manner based on the local farmers' knowledge and imagination.Strengthening the SIAL in the municipality of Morón on the basis of its principles, in order to contribute to its Food Security or Food Sovereignty, is a challenge that will provide a short, medium and long-term response to local agricultural development, will promote adequate articulation of the agents of change in the agricultural context, will mitigate the lack of perception among stakeholders about local development, as well as the lack of awareness among stakeholders of the need to build innovative capacities among producers, academics, scientists and decision-makers to continue with implementation at a more strengthened level.In view of the above, it is necessary to continue creating conditions that make possible the integration of actors in local development contexts in the municipality. The following **scientific problem arises** from these problems: How to contribute to Food Sovereignty in the municipality of Morón, with the **general objective of** designing a multi-stakeholder action plan based on the strengthening of the principles of SIAL to contribute to Food Sovereignty in the municipality of Morón.

Specific objectives:

1. Theoretical and methodological foundation of the principles of SIAL and Food Sovereignty.
2. To characterize the agricultural process in the Morón municipality of Ciego de Avila province.
3. To characterize the innovation context at the local level in the municipality of Morón.

4. Propose a multi-stakeholder action plan to strengthen the SIAL in the municipality of Morón.

The **scientific hypothesis** defended in this research is that if a multi-stakeholder action plan is designed from the strengthening of the principles of SIAL, it can contribute to Food Sovereignty in the municipality of Morón.The methodology used is based on the assumptions of qualitative research from the premises of a mixed approach and the sample used was purposive, proper in conducting an exploratory study (Table 1).

Table 1. Sample selected by sources of information.

Morón

Local Government 2

University 2

Agricultural Company 3

Unit Producers 7

Productive

Women inserted in the 8

productive units

Facilitators 4

The methodological conception that sustains the work is based on analysis-synthesis, induction-deduction and historical-logical. The analysis-synthesis is used in the criteria exercise. It facilitates an approach to the ideas approached for the study of the agricultural context and innovation in Morón, discovering their relationships and the general characteristics among them.The induction-deduction allows, based on the study of the Local Development Strategy and the interviews and visits to meetings, to reach generalizations and identify the main strengths and weaknesses of the agricultural context and local innovation in the municipality of Morón.On the other hand, the historical-logical method is necessary to approach the treatment of the SIAL from the scientific literature, as well as the historical evolution of the municipality of Morón, the local development strategy.Discussion groups were created with local government actors, the University of Ciego de Avila (UNICA), the productive sector, sampling, application and tabulation of instruments (interviews with women, young people, interviews with producers), which allowed corroborating the real state of the productive and innovative context of the municipality of Morón. In addition, we monitored the Facebook page "Morón Hoy", a historical and cultural page of the group of social communicators, with 5,000 friends residing in the municipality.For the processing of the information and the analysis of the quantitative data obtained from the statistical analysis for those recorded in sources such as those contemplated by the National Office of Statistics and Information (ONEI), those contemplated in the Land Census office in the Provincial Delegation of Agriculture in Ciego de Avila, in addition, the document on the "Integrated system of sustainability evaluation in Cuba was consulted. Case study: Morón municipality, province of Ciego de Avila. CITMA's Center for Capacity Building in Morón, for the evaluation of qualitative data an interpretation of the discourse was made. **The practical contribution is the** multi-stakeholder action plan to strengthen the principles of SIAL and thus contribute to Food Sovereignty in Morón. The municipality has some It also has the conditions to continue strengthening the implementation of the SIAL with effectiveness and encouraging results for the population.

DEVELOPMENT

1. THEORETICAL AND METHODOLOGICAL FOUNDATION OF THE PRINCIPLES OF SIAL AND FOOD SOVEREIGNTY.

The Local Agricultural Innovation Systems respond to the particular agroecological and productive characteristics of the territories, as an instance of consultation, coordination, planning, implementation, monitoring and evaluation of agricultural research and innovation processes. Ortiz, R., la O, M., and Miranda, S. (2017).On the other hand, Ortiz, R., la O, M., and Miranda, S. (2017) referencing Nuñez (2007), consider that much attention must be paid to the actors involved in innovation systems. The interrelationships between the actors must be an integrated whole, since it constitutes a system. In this regard, Fis, Y., Arzola, L., and González, K. (2019) emphasize the necessary participation of all the actors involved in the agricultural and local innovation contexts. The collective construction of the SIAL makes it possible to achieve changes in the territories, to have a positive impact on local development and the use of resources.For Ortiz Pérez, H., Acosta Roca, R., Ruz Reyes, R., Arias, M., Rivas Diéguez, A., & Núñez Jover, J. (2021) and Jover JN, Ortiz HR, Proenza T, Rivas A. (2020), the SIAL integrates components and participatory processes to respond to local demands. Where, Ortiz Pérez, H., Acosta Roca, R., Ruz Reyes, R., Arias, M., Rivas Diéguez, A., & Núñez Jover,J. (2021) state that from the horizontal construction of work strategies in correspondence with the challenges of the agricultural context, it has been possible to increase yields in a sustainable manner, through the efficient use of resources and energy. This proposal contributes to the generation of jobs, prioritizing gender and generational equity, while supporting the strengthening of responsible and participatory governance with the empowerment of producers and their cooperatives.Authors who propose the design of the SIAL organization, the innovation management cycles and the capacities to be created to apply the principles and good practices of the system. Figure 3.From the perspective of the actors themselves, it improves social welfare and promotes mass access to the diversity and knowledge of agrobiological, technological and social resources through action learning processes. Knowledge and technologies are connected with social needs/demands and a horizontal interaction of actors is achieved through the creation of participatory spaces; where dialogue, consensus building and common visions are part of the collective action that promotes the use of diversity in favor of food security and food sovereignty. In this way, agricultural systems are strengthened for adaptation and resilience to climate change; healthy, diversified and quality food is promoted; natural resources are managed in a sustainable manner and simple and practical mechanisms are implemented for disaster risk management at the local level. Ortiz Pérez, H., Acosta Roca, R., Ruz Reyes, R., Arias, M., Rivas Diéguez, A., & Núñez Jover, J. (2021)

Figure 3. General outline of the SIAL: organizational design, innovation management cycles and capacities to be created to apply the principles and good practices of the system.

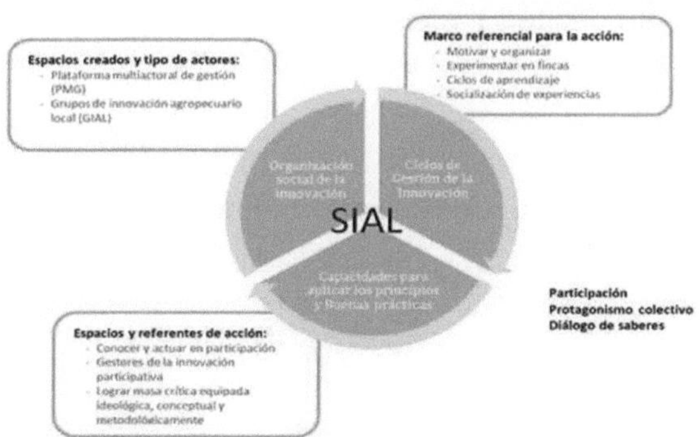

Source: Ortiz Pérez, H., Acosta Roca, R., Ruz Reyes, R., Arias, M., Rivas Diéguez, A., & Núñez Jover, J. (2021).

For these researchers, it is necessary to use a new methodological tool related to the design and implementation of learning cycles (Figure 4), where the social interaction generated among the actors and the promotion of innovation management are instruments for change and improvement of the localities involved.

Figure 4. Working approach used for SIAL implementation: action learning cycles.

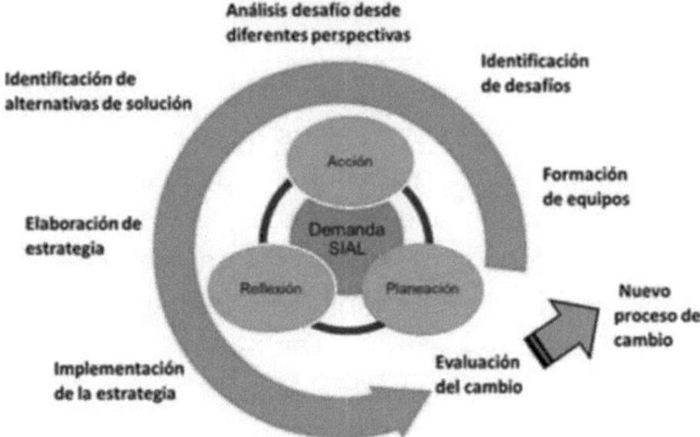

Source: Ortiz Pérez, H., Acosta Roca, R., Ruz Reyes, R., Arias, M., Rivas Diéguez, A., & Núñez Jover, J. (2021).

At this point it is significant to note that at the Forum on Food Security: Challenges of the Future and Tasks of the Present held in 2013, Fernando Eguren defined the components of Food Security as: Supply, Access, Use, Stability and Institutionality. Figure 5. **Food Security Forum (2013).**

Figure 5. Components of Food Security.

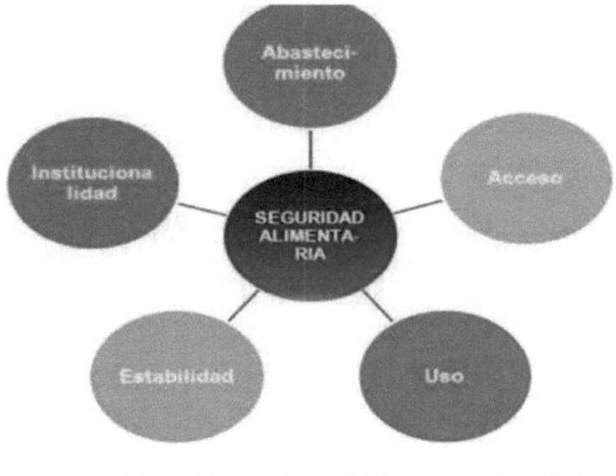

Source: Foro Seguridad Alimentaria, (2013)

Analyzing the components of Food Security or Food Sovereignty such as: Supply, Access, Use, Stability and Institutionality together with the principles of SIAL such as participation, collective protagonism and dialogue of knowledge, in the context of the municipality of Morón, brings with it this new and necessary paradigm of promoting innovation, and its implementation in its essence is an innovative act that requires learning in action and demands the participation of all local actors, accompanied by the academy, as shown in Figure 6.

Figure 6. Relationship between the components of Food Sovereignty and the principles of SIAL in the municipality of Morón.

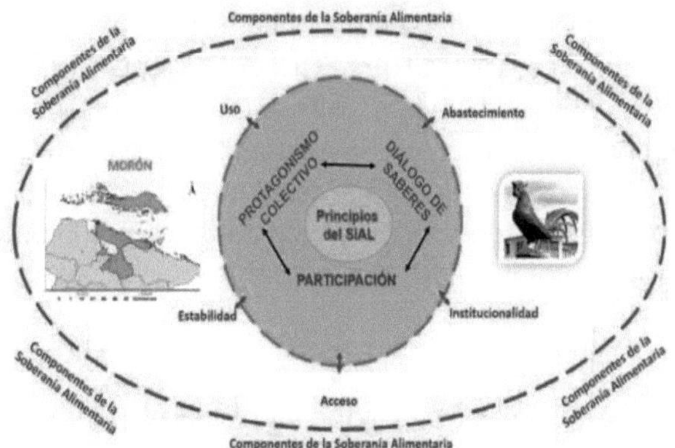

Source: Own elaboration.

It is significant to note that, in the Plan for Food Sovereignty and Nutritional Education in Cuba (2020), a diagnosis of Food Security and nutritional education in Cuba is made with extensive and detailed information, coordinated by the National Institute of Economic Research in the country, and that for the development of this research has been a material of constant consultation.

2. ANALYSIS OF THE AGRICULTURAL CONTEXT OF THE MUNICIPALITY OF MORÓN

In the municipality of Morón, from the point of view of agro-productivity of agricultural soils in the municipality, the most representative are category III with 30.6%. These values do not include the wetland, which has more than 30,000 ha, representing 52.8% of the total land area of the municipality that occupy marshy soils located in the lowlands, coastal areas and on the periphery of the lagoons with limited agricultural use. This situation means that the regions most exploited by agriculture are located towards the south of the municipality, where the most agriculturally productive soils are also located. The main land holders in the municipality are: MINTUR, AZCUBA, MINAG, MINAL, MININT, VIALES and human settlements, with MINAG being identified as the largest holder. Land use is composed of livestock, agriculture, conservation, forestry, water, human settlements, tourism, facilities and infrastructure, with conservation being the most common use, taking into account that the territory is located within the Great Northern Wetland. Figure 7

Figure 7. Soil types in the municipality of Morón.

Source: Municipal Statistics Office (OMEI) of Morón.

In Morón, land will continue to be an important resource in the municipality, so its rational use is fundamental in the development of the territory. A greater correspondence is achieved between the potential offered and its use, also taking into account the water factor. For this purpose, it is necessary to recover those soils that are idle and are of agrological category I and II, mainly in the southeastern and central portion (intensive agricultural zone) where water resources are available. This is shown in Table 1.

Table 1. Proposed Territorial Transformations for 2030

Indicator	2030
Agricultural land	17 495
Cultivated area	9 427,39
Surface Cult. Temporary	3 201,49
Sup. Cult. Permanent	4 192,1
Sugar Cane	3 338,43
Livestock	9 553,65
Forestry	56 023,98
Idle Surface	108

Source: Local development strategy of the municipality Morón.2020.

However, it is important to highlight that agricultural development is increasing, based on food production as a fundamental premise as one of the development strategies, Figure 8, since precise objectives are assumed in this regard, such as:

1. The recovery of more than 90% of idle land, which increases production areas by more than 250 ha for various crops and approximately 1,500 ha for livestock.
2. Development program for the Turiguanó Genetic Livestock Farm based on the recovery of the areas of the units and reaching its maximum load (0.97 UGM), in addition to recovering the areas of the units in disuse (according to program).

Figure 8. Agrological capacity of soils for 2030.

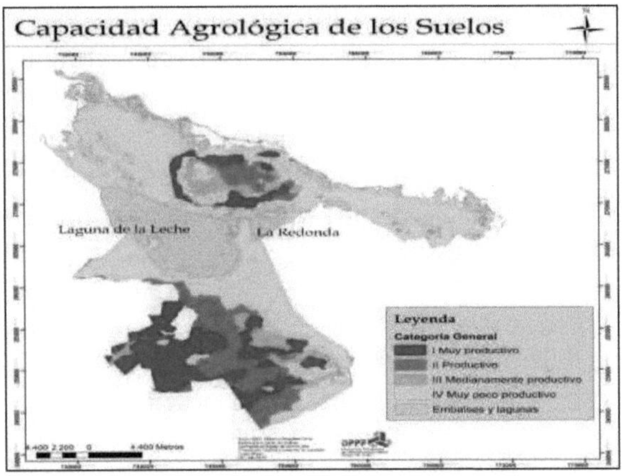

Source: Municipal Statistics Office (OMEI) of Morón.

The municipality has a high biodiversity represented by a flora with 46 species of vascular plants belonging to 25 families and three endemic species: Atkinsia cubensis (black Majagua); Pithecellobium obovale (Encinillo) and Copernicia gigas (Hediondo). The main values of use of the vegetation are given in that 34.78% of the species present have value as medicinal plants, 17.77% as food, 6.66% as toxic or poisonous, 17.7% as melliferous, 41.30% as ornamental, 13.33% as animal food, 56.52% as timber, and 28.26% of the total with other uses.Fauna is represented by 119 species of terrestrial invertebrates grouped in three classes, 12 orders and 36 families with one species endemic to Cuba, 102 species of insects, 15 species of arachnids and two species of terrestrial mollusks with one species endemic to Cuba. Terrestrial vertebrates have 98 species (nine reptiles, 86 birds and three mammals) with 22.4% endemism. Of the birds, 77 species are common in Cuba, three are endemic and five are endangered, such as the Parakeet (Amazona leucocephala).Morón has a large part of its areas covered by mangrove forests and vegetation associated with the Great Northern Wetland of Ciego de Avila, which has the category of Wetland of International Importance and was declared a Ramsar Site on November 18, 2002.The current forest area is 44,915.3 ha for a coverage of 41.36% of the patrimony. The main holders of these forests are the Empresa

Forestal Integral and the Empresa para la Protección de la Flora y la Fauna, as well as private sector farmers.There are a series of sites in the territory that have natural values that require special attention in the sense of protecting and recovering them, most of which are identified within the Protected Areas System, which includes the approved Protected Area, which is the Center and West of Cayo Coco, with the category of Ecological Reserve by Agreement No. 6803/2010 of the Executive Committee of the Council of Ministers; the Great Wetland of Northern Ciego de Avila, with recognition as Wetland of International Importance, RAMSAR Site according to Article 2.2.2. 6803/2010 of the Executive Committee of the Council of Ministers; the Great Wetland of the North of Ciego de Avila, with recognition as Wetland of International Importance, RAMSAR Site according to Article 2.1 of the Convention on Wetlands, registered in site 1235 since November 18, 2010. The area is proposed as a Managed Resource Protected Area, as well as two other proposed areas, which are the Playa Pilar Dunes on Guillermo Key as an Outstanding Natural Feature and the La Leche-La Redonda lagoon system as a Protected Natural Landscape.The Integrated Coastal Management Program of the Great Northern Wetland of Ciego de Avila, approved by the Ministry of Science, Technology and Environment in 2010, is being developed in the territory; its management body is a Coordinating Board that is integrated by all the economic and social actors that develop their activity in this important ecosystem. The Program's main lines of work include water and forest resource management, social communication and environmental education, research, monitoring, and biodiversity protection.It has 38 km of beaches for sun and beach tourism distributed by 17 beaches in the Jardines del Rey Tourist Destination, the coastal processes of the territory are regulated by the country's environmental legislation Decree Law 212 "Management of the coastal zone", site in which the action of the trade winds and therefore its effects on the waves is exacerbated. It has allowed the formation of numerous beaches and dunes, with the existence of a Recovery, Rehabilitation and Maintenance Program in coastal areas. In 2020, sand will be dumped on Cayo Paredón Grande on North Beach, and in 2021, sand will be dumped with an approved project on Middle Beach on Cayo Guillermo.The total population of the municipality is 70,595 inhabitants, residing in 19 concentrated settlements, of which 2 are urban and 17 are rural. There is also a dispersed population that in recent years has had an initial process of decrease and then tends to increase in locations that have remained in the closest settlements. Table 2

.

Table 2. Number of population and housing by settlement in the municipality of Morón.

Population and housing by settlements				
Settlements	Category	Housing	Population	People's Council
Morón City	Urban	20836	65 575	East, West, South, Homeland, and Vaquerito
Sandino-La Loma	Urban	947	2518	Turiguanó
El Salado	Rural	60	170	Turiguanó
Dutch	Rural	60	202	Turiguanó
Commander Manuel Fajardo	Rural	129	382	Turiguanó
San Rafael	Rural	49	147	Turiguanó
The Manatee	Rural	51	120	Turiguanó
The Ecil	Rural	32	84	Cowboy
Loma Ciega	Rural	81	225	West
El Quemado	Rural	60	161	Homeland
Camilo Cienfuegos	Rural	39	104	Homeland
El Chillante	Rural	70	217	Homeland
Santa Barbara	Rural	103	289	Homeland
La Teresa	Rural	32	84	Homeland
Saladriga	Rural	121	333	Homeland
The Rose	Rural	148	393	Homeland
La Serrana	Rural	79	217	Homeland
El Palmar	Rural	22	56	Homeland
Eden	Rural	68	203	Homeland
Grand Total		22987	70 595	

Source: Municipal Statistics Office (OMEI) of Morón (information prior to the 2012 census).

Morón exhibits a dynamic socioeconomic context with high levels of social mobility due to the incidence of two essential phenomena population aging with 17.2% of the population.The municipality is in aging group III, as is the country (according to the category established for the countries of the region by the Latin American and Caribbean Demographic Center of ECLAC), which shows the importance of inserting them more and more to the active life of the community and provide them with a better quality of life and internal migrations with a rate that shows a stable growth rate, closing 2019 with 449 inhabitants. The predominant pattern is the existence of negative balances in external migrations, while the internal migratory balance is positive, which shows that it is an area of attraction as a population receiving area. The working age population at the close of 2019 was 45 634, of them 23 624 males for 51.8 % and 22 010 for 48.2 %; in pre-working age was 945 total of them 504 were males and 441 females and in post-working age a total of 11.11 of them 4 230 males and 6 887 females. They are incorporated to work 13 472 moroneros that represents a percentage of unemployment of 29.52 %, if it is analyzed in addition that at the closing of the information in 2019 the relation of economic dependency of people of more than 65 years against the economically active population is of 73.61, reason why the projection of the municipality must be directed to incorporate to the labor activity those inhabitants that do not exercise any labor activity that supports to the growth of the population aging.The occupational structure in the municipality shows a stable behavior in terms of its general classification, the state sector has 415 administrative managers (171), 5346 technicians, 2893 service workers and 4647 operators for a total of 13,000 employees, and the state sector has 415 administrative managers, 5346 technicians, 2893 service workers and 4647 operators for a total of 13,000 employees.472. Self-employed workers 4 310, farmers 1 276 housewives 10 876, retirees 11,080, students 11,495. The service, agriculture and construction sectors are those that provide the greatest number of jobs in the municipality, which is why most of the population is linked to them. The city of Morón is one of the main employers of workers linked to tourism services in the northern keys, currently employing 3707 workers, 57.3% of the total number of workers in this sector. The rest are service workers, predominantly in the education,

health, commerce and industrial sectors in the urban settlements. The administrative service based in Morón is also representative.In the present investigation a study was carried out, Annex 1, on the valuation of the position of women in Morón, within the current agricultural context, from their roles, access to products and services, improvement, as well as in decision making, which was applied to a sample of 15 people where 7 are producers of the Productive Units and 8 are women inserted in the productive units and according to the type of work they do, women are dedicated 80% to support work in the units while men are the ones who mainly work on the land.On the positions they occupy and men are the most prevalent in the board of directors and important roles within the cooperative. In addition, the stimulation for the work that women develop in the productive unit is not taken into account. Regarding the cooperative's decisions and how women participate and in which ones men participate, women's participation in decision making is still not valued, which represents 10% that it is.And the practical needs of women are not addressed in relation to restrooms in production sites, clothes and shoes with the right sizes for them, work implements suitable for them, meeting schedules, among other issues.The planning of training activities and events is not taken into account, except by the Federation of Cuban Women in which women participate. And the main resources and inputs in the cooperative are controlled by men. In other words, in general, in Morón in the Productive Units, women's work is not adequately valued.It is important to know that in Morón, municipality of Ciego de Avila, the local development strategy has the **Mission of**:
Satisfy the economic and social needs of the population, by taking advantage of endogenous potential and exogenous resources with emphasis on tourism, with a rich historical and cultural heritage, where traditions and the beauty of the landscape join its hospitable people, proud of their land and committed to promoting sustainable development based on innovation and efficient use of resources, ensuring the protection of ecosystems and gender mainstreaming.And that the **Vision** is:Morón, Tourist Municipality, Capital of Jardines del Rey, a center of agricultural, industrial and artisan production, competitive for the quality of its services, where the feeling of belonging is conquered, an educating, scientific and innovative municipality, healthy and in harmony with the environment.

And that its strategic lines are:

1- Agricultural and forestry development for self-sufficiency and food sovereignty.
2- Innovation-based local tourism management

3- Development of industry, services, export lines and import substitution.
4- Infrastructure (sewage, roads, renewable energy) and production of construction materials.
5- Responsible environmental management for the conservation of the ecosystem (Task Vida)

6- Development of Information and Communication Technologies (ICTs) in favor of the informatization of society for an inclusive and sustainable development.

The municipality has 107 Legal Entities (40 State Enterprises, 34 of national subordination, 5 of provincial subordination and one of municipal subordination; 22 Budgeted Units, 9 of national subordination, 5 of provincial subordination and 8 of municipal subordination; 2 banking entities; 4 Agricultural Production Cooperatives; 6 Strengthened Credit and Service Cooperatives; 3 Basic Cooperative Production Units; 21 Totally Cuban Capital Mercantile Corporations; 4 Mixed enterprises; 3 Political and Mass Organizations; 1 Non-Profit Civil Society and 1 Non-Governmental Organization; 3 Basic Units of Cooperative Production; 21 Totally Cuban Capital Mercantile Companies; 4 Mixed enterprises; 3 Political and Mass Organizations; 1 Non-Profit Civil Society and 1 Non-Governmental Organization), and 2,605 TPCP Individuals, of which 201 are home renters, 648 are carriers and 1,756 are self-employed workers in other activities (of which 297 are contracted).In agriculture, livestock and forestry at the end of 2019, 773 workers were employed, including 229 women.An analysis was made of the data obtained (ONAI) from which it was deduced that in 2019 there was a waste of the agricultural area of the municipality (of the 17800 ha of agricultural area 8800ha was cultivated) that the state agricultural sector was the one that least used the area 9100ha/3800), in the non-state sector of 8700ha 5000ha were cultivated, the UBPC have the lowest use (2600ha/400ha) and the one that presented the best use were the CCS and private (3900ha/ 2700ha). A detailed analysis of the data obtained by ONAI related to harvested area and production of selected state and non-state non-cane crops in the last five years shows a considerable decrease, especially in banana, cassava, pumpkin, beans and total fruit. Table 3.The total number of agricultural hectares in the municipality is 2,871 cultivable among all producers, for the demand of the municipality it is necessary to take full advantage of the entire agricultural area and the construction of organoponics on non-agricultural land.

Table 3. FOOD BALANCE BETWEEN DEMAND AND SUPPLY IN THE MUNICIPALITY 2020

YEA R S	INDICAT OR EN	um	SUPPLY-DEMAND BALANCE						
			TOTA L	VIAND A S	HORTAL IZA S	FRUIT S	GRAIN S	MIL K	CA RN E
Year 2021 Plan	Sowing	Ha	750.8	154.6	170.7	9.0	416.5	410000 L	47 T
	Productio n	t	3955. 7	2543.6	1154.2	158.5	99.4		
	Yield	t/h at	41.05	16.45	6.76	17.61	0.23		
	Demand	t							
	% Satisf.	%							
Actu al Year-end 2020	Sowing	has	1362. 0 4	357	312.14	18.5	674.4	238597 L	
	Productio n	ton	4270	1898.9	1691	275.5	404.6	253835 L	
	Yield	t/h at	26.21	5.31	5.41	14.9	0.59	106 %	
	Demand	T							
	% Satisf.	%							

Note:

- In the spring 2020 campaign, 548.56 ha were planted.
- In the current 2020-2021 cold season, 496.34 ha are forecast to be planted to date 273.96 ha have been planted.

Source: Municipal Statistics Office (OMEI) of Morón (information prior to the 2012 census).

In the current situation, the municipality cannot meet the 30lb per capita per inhabitant because there is very little agricultural land for the amount of population. At this time, we are contracting with all the farmers to commit them to plant 100% of the agricultural land. The UEB urban farm (it is of provincial subordination and belongs to the Cítricos Ceballos Company) its productions are stockpiled for Morón, milk with Morón dairy and for commerce and gastronomy.The Sergio Gonzalez CCS process is currently being analyzed because there are 68 deceased landowners and the corresponding legal processes have not been carried out in each case, which prevents the proper use of the land and the contracting of agricultural production.The Urban Farm Company with its respective production bases is planning to export its products through the Ceballos Agroindustrial Company. Through this company, Morón's producers will be linked to the tourism sector.This Urban Farm Company to contribute to the development of the municipality is projected in the planting of aloe vera and natural mango for export, also for planting, processing and marketing of vegetables and fresh vegetables. Including the projects of DL in the organoponic Patria 2 in phase of confection of the project.The Granja Urbana Company produces organic matter for productive improvement with solid and liquid worm humus and compost that produce 125 T using agricultural waste.Although there is no reliable information to determine as accurately as possible the potential of the municipality, in this sense it is analyzed that, in the municipality, neither the company GELMA de Suministro Agropecuario, nor the Delegation of Agriculture are projected for the development of exports and their productive chaining, nor are they projected for the export of services, neither in scientific advice nor in training. They do not have the technical and trained personnel for this.In the year 2020, Gelma did not project in its initial plan the income from export of goods and services, nor expenses for the cost of export of services, nor project expenses, nor event expenses, thus limiting the possibility of incorporating science and research to the increase and improvement of its management, nor financial income from financing received from companies, nor from donations received. We cannot explain the causes of these non-compliance indicators that affect the Food Sovereignty plan. Summary of the main weaknesses from the local development strategy of the municipal government:

- Lack of awareness, knowledge and experience in the agricultural sector regarding exports and production linkages.
- Little development of the agroindustrial sector.
- Insufficient strategic management of managers for decision making.
- The land balance is not updated.
- Existence of land that remains idle or poorly exploited.
- No alternatives are applied to expedite the delivery of land for immediate exploitation to the applicants.
- Deficiencies in the agricultural sector management model.
- There are no swine agreements in the municipality.
- Insufficient small livestock and poultry breeding stock.

2.1 SWOT matrix of the agricultural context to strengthen SIAL in the municipality of Morón.

For the determination of the SWOT Matrix we followed the steps shown in Figure 9 proposed in 2012 by Dr. Renata Marciniak when discussing strategies, models, management tools and other information necessary to know how to manage a company. Marciniak, R (2012)

Figure 9. Steps to follow for the construction of a SWOT Matrix.

Source: Marciniak, R (2012)

The assessment of all the information previously analyzed facilitated the identification of the local agricultural context in terms of strengths, weaknesses, threats and opportunities:

STRENGTHS:

• There is economic and agricultural development in the municipality.

• Integration between academia and research centers such as CITMA and CIBA and productive sectors.

• Existence of local enterprises based on the knowledge of producers of different grain crops and with involvement in the agricultural context.

• Highly qualified professionals are sensitized to act directly in local development programs.

WEAKNESSES:

• Little knowledge of the sources of financing for local development, with emphasis on agriculture and livestock.

• The capacity building that takes place is based on organizational forms that are not contextualized.

• Deficient involvement and use of agroecological techniques to achieve sustainable agriculture.

• Little use is made of the agro-productive potential to promote the design of local development projects.

OPPORTUNITIES:

• Change in policies that favor agricultural development at the country level.The reopening of Tourism after a prolonged period of stoppage.

• The consistency and excellent management of the local Agricultural Innovation System project in Ciego de Avila Province.

• The existence of a. Cuba's Food Sovereignty and Nutrition Education Plan. THREATS:

• The existence of an Integrated Framework for the classification of Food Security and the humanitarian phase.

• The existence of a pandemic that is currently sweeping the world.

• The demand for fresh and organic food from tourists visiting the hotels in the keys and the City of Morón.

Figure 10 below shows the relationships established in the agricultural sector between strengths, weaknesses, threats and opportunities, from the internal and external analysis, as well as the positive and negative aspects of each aspect mentioned.

Figure 10. Agricultural context for strengthening the SIAL in the municipality of Morón

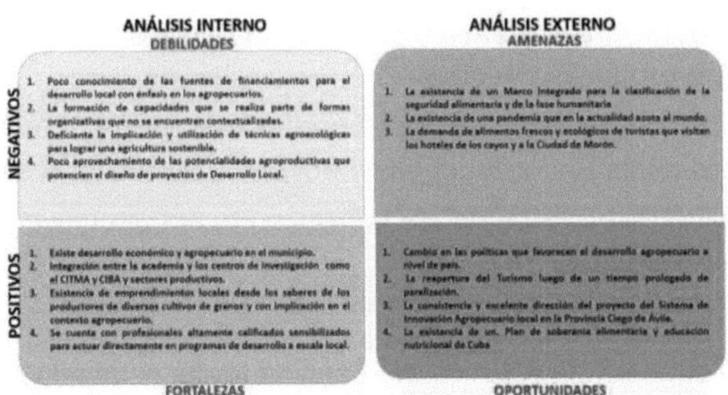

Source: Own elaboration.

As a result, the SWOT matrix of the agricultural context can be constructed, with its four strategies represented as follows: FO Maxi-Maxi STRATEGY:

Provide spaces for coordination between the municipal government, the CUM, producers, CITMA, CIBA and other stakeholders to plan actions to strengthen local agricultural development, taking into account the principles of the SIAL and actions of the country's Food Security Plan adjusted to the municipality of Morón.

FA Maxi-Mini STRATEGY:

To take advantage of the existence in the territory of local enterprises and the availability of highly qualified professionals to meet the demand for fresh and organic food from tourists visiting the hotels in the keys and the city of Morón, as well as to establish indicators related to the Integrated Framework for the classification of Food Security and the humanitarian phase and thus contribute to mitigate the consequences of the Covid-19 pandemic on the population of the municipality.

STRATEGY DO Mini-Maxi:

Assume the policies that favor agricultural development at the country level, and take advantage of the reopening of tourism for the sale of agricultural products, and thus continue to strengthen the SIAL in the territory by applying its principles to achieve a more sustainable development. Food Security, and to train facilitators to multiply agroecological techniques and different forms of financing for local development projects.

DA Mini-Mini STRATEGY:

Work on capacity building in the context and strengthening of SIAL by applying its principles and providing knowledge of sources of financing for local development with emphasis on agriculture and agroecological techniques for sustainable agriculture to meet the demand for fresh and organic food for tourists visiting the hotels in the keys and the city of Morón.The relationships established between the strengths, weaknesses, threats and opportunities and the four strategies can now be represented, thus forming the SWOT matrix of the agricultural context for strengthening the SIAL in the municipality of Morón. Figure 11.

Figure 11. SWOT matrix of the agricultural context to strengthen the SIAL in the municipality of Morón.

Contexto *AGROPECUARIO* para fortalecer el SIAL en el municipio Morón

MATRIZ FODA	FORTALEZAS (Optimizarlas)	DEBILIDADES (Minimizarlas)
	1. Existe desarrollo económico y agropecuario en el municipio. 2. Integración entre la academia y los centros de investigación como el CITMA y CIBA y sectores productivos. 3. Existencia de emprendimientos locales desde los saberes de los productores de diversos cultivos de granos y con implicación en el contexto agropecuario. 4. Se cuenta con profesionales altamente calificados sensibilizados para actuar directamente en programas de desarrollo a escala local.	1. Poco conocimiento de las fuentes de financiamientos para el desarrollo local con énfasis en los agropecuarios. 2. La formación de capacidades que se realiza parte de formas organizativas que no se encuentran contextualizadas. 3. Deficiente la implicación y utilización de técnicas agroecológicas para lograr una agricultura sostenible. 4. Poco aprovechamiento de las potencialidades agroproductivas que potencian el diseño de proyectos de Desarrollo Local.
OPORTUNIDADES (Aprovecharlas) 1. Cambio en las políticas que favorecen el desarrollo agropecuario a nivel de país. 2. La reapertura del Turismo luego de un tiempo prologado de paralización. 3. La consistencia y excelente dirección del proyecto del Sistema de Innovación Agropecuario local en la Provincia Ciego de Ávila. 4. La existencia de un Plan de soberanía alimentaria y educación nutricional de Cuba.	**ESTRATEGIA: FO (MAXI-MAXI)** Proporcionar espacios de concertación para la articulación entre el Gobierno Municipal, el CUM, los productores, el CITMA, CIBA y otros actores en aras de planificar acciones para fortalecer el desarrollo agropecuario local, asumiendo los principios del SIAL y acciones del Plan de seguridad alimentaria del país ajustadas al municipio Morón.	**ESTRATEGIA: DO (MINI-MAXI)** Asumir las políticas que a nivel de país favorecen el desarrollo agropecuario, y aprovechar la reapertura del turismo para la venta de productos agropecuarios, y con ello seguir fortaleciendo el SIAL en el territorio aplicando sus principios para lograr una seguridad alimentaria, y formar facilitadores que multipliquen las técnicas agroecológicas y las diferentes formas de financiamiento para proyectos de desarrollo local.
AMENAZAS (Superarlas) 1. La existencia de un Marco Integrado para la clasificación de la seguridad alimentaria y de la fase humanitaria. 2. La existencia de una pandemia que en la actualidad azota al mundo. 3. La demanda de alimentos frescos y ecológicos de turistas que visitan los hoteles de los cayos y a la ciudad de Morón.	**ESTRATEGIA: FA (MAXI-MINI)** Aprovechar la existencia en el territorio de emprendimientos locales y la disposición de profesionales altamente calificados para atender la demandas de alimentos frescos y ecológicos de turistas que visitan los hoteles de los cayos y a la Ciudad de Morón además establecer indicadores relacionados con el Marco integrado para la clasificación de la seguridad alimentaria y de la fase humanitaria y así contribuir a atenuar las secuelas que ha dejado la pandemia de la Covid-19 a la población del municipio.	**ESTRATEGIA: DA (MINI-MINI)** Trabajar en función de la formación de capacidades en la contextualidad y en el fortalecimiento del SIAL aplicando sus principios y brindar conocimientos de las fuentes de financiamientos para el desarrollo local con énfasis en los agropecuario y en sus técnicas agroecológicas para una agricultura sostenible al atender las demandas de alimentos frescos y ecológicos de turistas que visitan los hoteles de los cayos y a la ciudad de Morón.

Source: Own elaboration.

Next, the same analysis is carried out in the context of innovation to then propose a plan of actions to be developed in the municipality of Morón and that in some way responds to the implementation of the 8 strategies of the SWOT matrices, which responds to the objective of the research, by assuming the chronological steps to achieve the creation and strengthening of the SIAL, proposed by Ortiz. R, Acosta. R, Angarica. L and Guevara. F, (2017).

3. CONTEXT OF INNOVATION AT THE LOCAL LEVEL IN THE MUNICIPALITY OF MORÓN

Innovation is a social process that involves participation and interrelation of people or groups and its materialization requires the indispensable presence of capacities to innovate, starting from the acquisition of existing knowledge, its use and assimilation, to the generation of new knowledge. It is the result of a process of social appropriation of knowledge (Arocena & Sutz, 2006), (Lage A., 2013) and (Nuñez, 2014).

Knowledge management for innovation consists of collaborating with the identification of local problems requiring knowledge, contributing with the organizations or individuals that can provide it, as well as building the links, networks and flows that allow the assimilation, evaluation, processing and use of this knowledge (Núñez, 2014). The local government of any territory must take into account the promotion of science, technology and innovation as pillars of its socioeconomic progress; the results of this triad must be committed to the welfare of its inhabitants, the eradication of inequalities and in harmony with nature.

In the municipality of Morón, a set of "potentialities" were identified that characterize it and make it one of the most important municipalities in the province of Ciego de Avila.The variety and quality of its natural landscapes and excellent beaches that are distributed in the northern part of the municipality, mainly in areas of the Great Northern Wetlands of Ciego de Avila, allow the increase of recreational and leisure activities ecologically well conceived and adaptable to the existing natural environment, both in the keys that already have their development plan, as in the area of the Embarcadero - Laguna de la Leche and Laguna la Redonda.The forest resource has a vast representation in extension and variety with endemic and exotic species. There are areas of excellent beauty for the development of trails, such as the approved and newly proposed Protected Areas, particularly in the northern part of the municipality. High potential for water resources due to the accumulation of surface water in the natural lagoons and the richness of the subway aquifer, with great potential for recreational activities, sports and lodging in the Laguna de La Leche and La Redonda, as well as to promote aquaculture.Due to its geographical location, the city of Morón is the main transportation hub for the entire northern area of the province, where most of the provincial automobile and railroad transportation routes converge.Presence of higher level services (Provincial Hospital, FORMATUR, Medical Sciences, Schools, Polytechnics, Municipal University Center and other specialties), which have a provincial character given its category of being the center of the Northern Region of the Province.In addition, the Centro Universitario Municipal CUM has a wide participation in networks (Red GUCID, Red CYTED, Red Académica "Diálogos en MERCOSUR", Red Iberoamericana de Docentes. Spain, Social Network for Researchers and Disseminators DivulgaRed. Fundación Andaluza para la Divulgación de la Innovación y el conocimiento, Red de Docentes de América Latina y el Caribe, RedDOLAC. Mexico, Red Nacional de Pequeños Rumiantes de Investigación - Producción (CIMAGT) and Red Nacional de Estudios de Población. CEDEM, the Federación de Ovejeros y Cabreros de Latinoamérica (FOCAL), the Grupo de Estudios Penales Internacionales, at the Universidad de Juan Carlos, Madrid, Spain, Grupo del Instituto de Derecho Internacional, Spain, the Confederación Panamericana de Escuelas de Hotelería y Turismo (CONPEHT), the Cuban Society of International Law of the National Union of Jurists of Cuba (Vice-Presidency) (Union of Jurists) and the Cuban Society of Administrative Constitutional Law (Union of Jurists), the Association of Pedagogues of Cuba, the Cuban Society of Psychology and ANEC, both national and international, in close liaison with research centers in the territory (CIBA, CITMA and CIE).And it participates in the socialization of science and innovation through publications in scientific journals and books and participation in national and international events. In order to continue strengthening the SIAL in the territory, a survey was made of the strengths, weaknesses, threats, and opportunities of the current context of local innovation in Morón.

3.1 - SWOT matrix of the innovation context to strengthen the SIAL in the municipality of Morón.

For the construction of the SWOT matrix, discussions were held with different local stakeholders (see Annexes 2 and 3) in the discussion groups at the "El Vaquerito" production unit located in the municipality of Morón, which is the unit of analysis for inserting and strengthening the SIAL, where, together with other stakeholder criteria, the strengths, weaknesses, threats, and opportunities of the existing innovation systems were obtained, as follows:

STRENGTHS:

1. The CUM's Science and Technology plan is in line with the territory's agricultural development priorities.
2. The University and the CUM are perceived as essential actors in the spaces of exchange, innovation and development.
3. Local stakeholders interested in promoting local development based on their potential.

WEAKNESSES:

1. Insufficient appreciation of the principles of local development in addressing local stakeholder linkages.
2. Lack of Local Development Projects that have in their conception the development of the locality.
3. Lack of a multi-stakeholder approach in innovation systems, which leads to low exploitation of S&T results at the municipal level.

OPPORTUNITIES:

1. Change in policies that favor the development of science, technology and innovation at the country level.
2. The reopening of tourism and other activities in the municipality after a prolonged period of paralysis.
3. The consistency and excellent management of the local Agricultural Innovation System project in Ciego de Avila Province.

THREATS:

1. The existence of a pandemic that is currently sweeping the world.
2. Lack of means for the development of agriculture and dependence on foreign management.

3. Practitioners' limitations in publishing their scientific results

Figure 12 below shows the relationships established in the context of innovation, between strengths, weaknesses, threats and opportunities, from the internal and external analysis, as well as the positive and negative aspects of each aspect mentioned.

Figure 12. Context of innovation to strengthen the SIAL in the municipality of Morón.

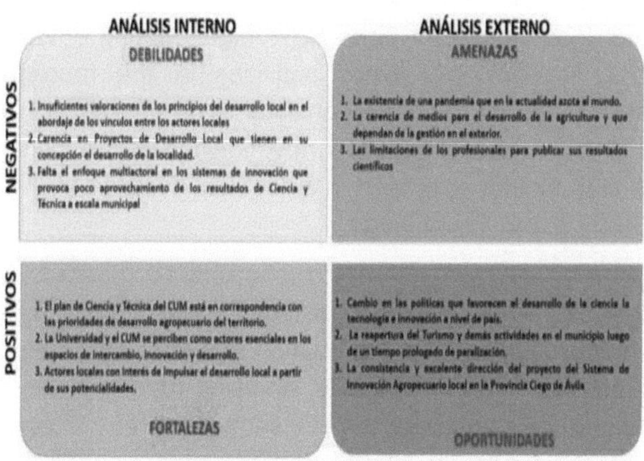

Source: Own elaboration.

As a result, the SWOT matrix of the innovation context with its four strategies can be constructed as follows:
FO Maxi-Maxi STRATEGY:

Provide spaces for coordination between the Municipal Government, the CUM, producers, CITMA, CIBA and other stakeholders to plan actions. to achieve and intensify local agricultural development, taking advantage of the new science policies in the country, the reopening of tourism and other activities in the municipality after a long period of paralysis, in addition to the strengths of SIAL in the province and the application of its principles to achieve adequate Food Sovereignty.

FA Maxi-Mini STRATEGY:

To take advantage of the presence of the University in the territory on behalf of the CUM, as essential actors in the spaces of exchange, innovation and development, in addition to local actors with interest in promoting local development to face the challenges of the pandemic and the lack of means for agriculture looking for alternative ways, in addition to searching for Databases of journals to publish scientific results.

STRATEGY DO Mini-Maxi:

Work according to the principles of local development in addressing the links between local actors for Food Sovereignty taking advantage of the reopening of Tourism and other activities in the municipality after mitigating the pandemic and the consistency and excellent direction of the project of the local Agricultural Innovation System in the Province for a new multi-stakeholder approach in innovation systems in the use of the results of Science and Technology.

DA Mini-Mini STRATEGY:

To strengthen the work in search of local development projects with a multi-stakeholder approach in innovation systems that respond to the Food Sovereignty of the municipality, applying the principles of SIAL, in order to mitigate in some way the consequences of the pandemic and the lack of means for the development of research in agriculture and the limitations to publish scientific results.The relationships established between the strengths, weaknesses, threats and opportunities and the four strategies can now be represented, thus forming the SWOT matrix of the agricultural context for strengthening the SIAL in the municipality of Morón. Figure 13.

Figure 13. SWOT matrix of the agricultural context for strengthening the SIAL in the municipality of Morón.

Contexto de INNOVACIÓN para fortalecer el SIAL en el municipio Morón

MATRIZ FODA	FORTALEZAS (Optimizarlas)	DEBILIDADES (Minimizarlas)
	1. El plan de Ciencia y Técnica del CUM está en correspondencia con las prioridades de desarrollo agropecuario del territorio. 2. La Universidad y el CUM se perciben como actores esenciales en los espacios de intercambio, innovación y desarrollo. 3. Actores locales con interés de impulsar el desarrollo local a partir de sus potencialidades.	1. Insuficientes valoraciones de los principios del desarrollo local en el abordaje de los vínculos entre los actores locales. 2. Carencia en Proyectos de Desarrollo Local que tienen en su concepción el desarrollo de la localidad. 3. Falta el enfoque multiactoral en los sistemas de innovación que provoca poco aprovechamiento de los resultados de Ciencia y Técnica a escala municipal.
OPORTUNIDADES (Aprovecharlas) 1. Cambio en las políticas que favorecen el desarrollo de la ciencia la tecnología e innovación a nivel de país. 2. La reapertura del Turismo y demás actividades en el municipio luego de un tiempo prologado de paralización. 3. La consistencia y excelente dirección del proyecto del Sistema de Innovación Agropecuario local en la Provincia Ciego de Ávila.	**ESTRATEGIA: FO (MAXI-MAXI)** Proporcionar espacios de concertación para la articulación entre el Gobierno Municipal, el CUM, los productores, el CITMA, CIBA y otros actores en aras de planificar acciones para alcanzar e intensificar el desarrollo agropecuario local, aprovechando las nuevas políticas de ciencia en el país, la reapertura del Turismo y demás actividades en el municipio luego de un tiempo prologado de paralización además de las fortalezas del SIAL en la provincia y la aplicación de sus principios para lograr una adecuada soberanía alimentaria.	**ESTRATEGIA: DO (MINI-MAXI)** Trabajar en función de los principios del desarrollo local en el abordaje de los vínculos entre los actores locales para la soberanía alimentaria aprovechando la reapertura del Turismo y demás actividades en el municipio luego de atenuar la pandemia y la consistencia y excelente dirección del proyecto del Sistema de Innovación Agropecuario local en la Provincia para un nuevo enfoque multiactoral en los sistemas de innovación en el aprovechamiento de los resultados de la Ciencia y Técnica.
AMENAZAS (Superarlas) 1. La existencia de una pandemia que en la actualidad azota al mundo. 2. La carencia de medios para el desarrollo de la agricultura y que dependan de la gestión en el exterior. 3. Las limitaciones de los profesionales para publicar sus resultados científicos.	**ESTRATEGIA: FA (MAXI-MINI)** Aprovechar la presencia de la Universidad en el territorio en nombre del CUM, como actores esenciales en los espacios de intercambio, innovación y desarrollo, además de los actores locales con interés de impulsar el desarrollo local para enfrentar los desafíos de la pandemia y de la carencia de medios para la agricultura buscando formas alternativas, además de buscar Bases de Datos de revistas para publicar los resultados científicos.	**ESTRATEGIA: DA (MINI-MINI)** Potenciar el trabajo en busca de proyectos de desarrollo local con enfoque multiactoral en los sistemas de innovación que respondan a la soberanía alimentaria del municipio, aplicando los principios del SIAL, para atenuar de alguna forma las secuelas de la pandemia y la carencia de medios para el desarrollo de investigaciones en la agricultura y las limitaciones para publicar los resultados científicos.

Source: Own elaboration.

The identification and use of resources and potentialities are pillars for the socio-economic development of a territory. The transformations that originate cross-cut the economic and social, generating an articulation and coordination between all local socioeconomic agents, public and private, all aimed at solving the existing difficulties and challenges in order to seek improvements in the living conditions of the inhabitants from the potentialities and the most efficient and sustainable use of local resources, which permeates it with the promotion of the capacities of local enterprises and the dynamization of the innovative space of the same.Next, a multi-stakeholder action plan is proposed to be developed in the municipality of Morón for the year 2022, which responds to the implementation of the eight strategies of the SWOT matrices, from the strengthening of the principles of the SIAL to contribute to Food Sovereignty, by assuming the chronological steps to achieve the creation and strengthening of the SIAL, proposed by Ortiz. R, Acosta. R, Angarica. L and Guevara. F, (2017).

4. MULTIFACTORIAL ACTION PLAN TO STRENGTHEN SIAL IN THE MUNICIPALITY OF MORÓN

Local development as a concept is at the center of controversies with a varied theoretical production, however, the political and ideological affiliation of the authors and the interests they represent, (Coraggio, 2002), (Vazquez, 2002), (Alburqueque, 2003) and (Arocena J., 1995) In most of them the economic component prevails.

For (Palma, 2006). Local development should promote the strengthening of urban structures, the local social business fabric, the use of available endogenous resources, the elimination of territorial inequalities and the mobilization and active participation of the citizen, through new participatory formulas in the political, social and economic spheres."...

For Cuba's conditions, local development is a perspective that has been present in the country's development practices, thought of as an integral process. Consequently, a Cuban conception of local development should be focused on equity, towards overcoming inequalities both territorial and between social groups..."(Labrada Silva, 2008)

The SIALs should respond to the particular agroecological and productive characteristics of the territories as an instance of consultation, coordination, planning, implementation, monitoring and evaluation of the processes of agricultural research and innovation.The SIAL cannot be conceived without the protagonism and activism of farmers and all the local stakeholders involved.

The culture of participation is an essential and determining component of the system. Human capital corresponds to the capacities of individuals to manage their demands for development in the collective scenarios of the system. It integrates conventional scientific knowledge and the traditional knowledge of the peasant imaginary.For the implementation of the SIAL, the culture of local and participatory innovation must be present, with at least one core team acting as a driving force in the territory.

This will be the auxiliary SIAL facilitation team. In practice, the rest of the local stakeholders must acquire the capacity to demonstrate the good practices that exist in each agricultural context and that this will lead to the creation of attitudes that will help them to develop a culture of innovation. innovative initiatives leading to a local development that is distinguished by its creativity and identity. In order to continue strengthening the SIAL in the municipality of Morón, it is necessary to follow the chronological steps proposed by Ortiz. R, Acosta. R, Angarica. L and Guevara. F, (2017). Figure 14.

Figure 14. Staggered chronological steps to achieve the creation and consolidation of the SIAL.

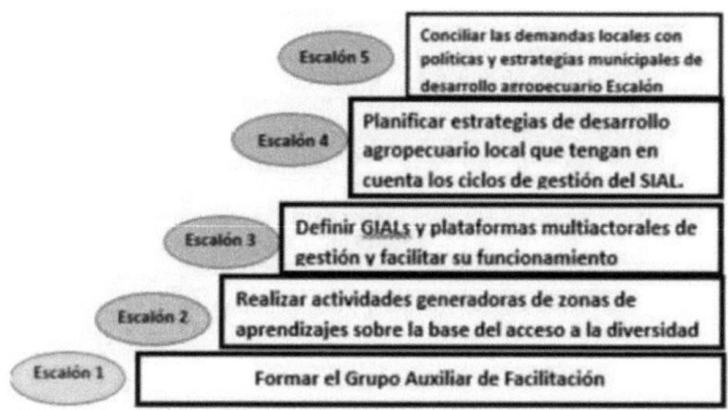

Source: Ortiz. R, Acosta. R, Angarica. L and Guevara. F, (2017)

Contextualizing these contents to the reality of the municipality of Morón, from the eight SWOT matrices constructed at the macro and micro social level, the following multi-stakeholder action plan is proposed, as a way to strengthen the SIAL from participation, collective protagonism and dialogue of knowledge, as the key to success of an agricultural development that focuses on equity and sustainability to contribute to Food Security.The following action plan is then presented based on the established approaches required for the integration of local stakeholders:

Table 6. Multi-sectoral actions to strengthen the SIAL in the municipality of Morón. Personal elaboration

Step 1	Shares	Responsible	Resources to be used	Indicators	Date
Form the Auxiliary Facilitation Group.	Dispatches with political and government authorities.	CAM y UNICA	Communication products Printed material, audiovisuals, tutorials and documents legal.	-Development of plans for Business. -Articulation of actors. -Conflict management. -Management skills. -Teamwork. -Knowledge of the actors and the context and its structures. -Level of participation - Collective leadership -Dialogue of knowledge -Existence of meeting minutes. -Knowledge of SIAL in the municipal context. -Existence of the group of facilitators and catalysts. -Formation and -Number of productive Bases.	January 2022
	Exchange and awareness meetings with PMG, GDL, CAM.	CAM y UNICA	Delivery of complementary communicative products, support material, etc.		
	Socialization of SIAL in the municipal and provincial context.	UNICA, TV avileña, ETECSA, Municipal government.	Information note Television broadcasts and promotional Spott. Local and provincial media (Radio		
	Exchange and reflection among, local development group, CAM, CUM, BANDEC, Centro municipal agriculture and their production units.	CAM y UNICA	Communication products Printed, audiovisual, tutorials.		
	Socialization and dissemination of SIAL through political mass organizations.	Political and mass organizations, CUM, UNICA.	Printed and audiovisual communication products, tutorials, articles, etc. scientists		February 2022
	Application of research instruments and techniques.	Organizations at political masses, CUM.	Surveys, interviews, focus groups, group and women's work, and		

			young people		
	Formation of a group of facilitators and catalysts that favor the insertion of the SIAL.	Municipal Government and UNICA			
	Follow-up and evaluation of the Diploma by the Coordination team.	UNICA	Group dynamics and satisfaction surveys.		
	Diploma Course: Local Agricultural Innovation System: For a participatory approach to management development.	UNICA	Courses or modules.		
Step 2	Shares	Responsible parties	Resources to be used	Indicators	Date
Carry out activities that generate learning zones. of learning.	Diagnostics for identify the needs, opportunities, problems, potential, demands, challenges and possible solutions.	Groupof facilitators and catalysts, UNICA, CAM.	Surveys, interviews and focus groups.	Level of participation Collective leadership -Dialogue of knowledge -Development of plans for Business. -Articulation of actors. -Conflict management. -Management skills. -Teamwork. -Knowledge of the actors and the context and its structures. -Number of facilitators. -Number of male and female producers.	March and April 2022
	Identification of actors, leaders and experiences of the different productive bases.	Group of facilitators and catalysts ,UNICA, CAM	Surveys, interviews and focus groups.		
	Creation of spaces for the exchange of debates, reflections for the conformation and formation of learning zones.	Group of facilitators and catalysts, UNICA, CAM.	Agroecological fairs and festivals, agricultural interest circle, farm visits, project creation, rural women's		

			movement, young peasant movement, ect.		
	Socialization of the results obtained in the learning zones created.	Group of facilitators and catalysts, UNICA, CAM, local and provincial media.	Communication products, brochures, pamphlets and scientific articles.		
Step 3	Shares	Responsible parties	Resources to be used	Indicators	Date
DefineGIALs /people's councilsand PMG and facilitate their operation	Formation of GIALs according to the different activities carried out.	Group f facilitators and catalystss,UNICA, CAM.	Surveys, interviews and focus groups.	Level of participation Collective leadership -Dialogue of knowledge -Development of plans for Business. -Articulation of actors. -Conflict management. -Management skills. -Teamwork. -Knowledge of the actors and the context and its structures. -Number of GIALs -Number of claims. -Number of farms. -Number of leaders.	May, June and August 2022
	Identification of interests and demands of producers for each GIAL.	Municipal Government and UNICA,	Participatory workshops, dialogue of knowledge with farmers and producers.		
	Mapping of the location of farms and leaders according to the GIALs formed with interest in contributing to development local.	group facilitation group, GIAL y UNICA	Agrodiversity fair.		

	Socialization of good practices based on the GIALs	group facilitation group, GIAL y UNICA	Communicative product, brochures, articles, participation in events.	-Number of consultation meetings. -Number of interactive workshops on best practices.	
	Coordination of exchange spaces between producers in farms and production units and research centers.	group facilitation group, GIAL y UNICA	Interactive workshop on best practices.		
	Collective construction and design of the platform's participatory action plan and annual plan.	Municipal Government, UNICA,y facilitation group and GIAL	Meetings of concertation.		
	Knowledge management based on the interests and needs of GIAL producers.	Groupfacilitation group, UNICA and GIAL	Information research, communicative products, talks, meetings with specialists, producers or leaders, school of campesino(a) s, micro-scholarships.		
	Diagnosis for the development of new knowledge.	Groupfacilitation group, UNICA y GIAL	Surveys, interviews, focus groups.		September and October 2022
	Elaboration of projects for local development	Municipal government, FacilitationFacilitation Group, UNICAand GIAL	Calls for proposals for local development projects, sources of financing for local development projects.		

Step 4	Shares	Responsible parties	Resources to be used	Indicators	Date
Plan local agricultural development strategies that take into account into account the SIAL management cycles	Motivation, awareness-raising workshops for local stakeholders.	Municipal Government, Facilitation Group, UNICA and GIAL	Communicative products, exhibition of results, fairs, contests, workshops.	-Level of participation -Collective protagonism -Dialogue of knowledge -Development of business plans. -Articulation of actors. -Conflict management. -Management skills. -Teamwork. -Knowledge of the actors and the context and its structures. -Number of GIALs -Number of claims. -Number of farms. -Number of leaders.	November 2022
	Diagnosis of the productive, social and environmental environment.	Municipal government, facilitation group, UNICA and GIAL	Survey, interview and focus group.		
	Mapping of the levels of perception of problems, potentials and demands.	Municipal government, Facilitation group, UNICA and GIAL	Survey, interview and focus group.	-Number of consultation meetings. -Number of interactive best practices workshops.	
	Design of action modalities.	Groupfacilitation group, UNICAand GIAL	workshops, NOPS techniques; discussion groups.		
	Socializationof the functioning and results and results of the GIALs to the municipal context.	Mediamedialocaland provincial media (Radio Morón, Radio Surco elInvasor, Punto Program on theIES ETECSA).	Information note, television and radio spots,		

Step 5	Shares	Responsible parties	Resources to be used	Indicators	Date
	Identify in the CAM and MINAG working group the spaces that provide opportunities to identify demands (sensitization of stakeholders - CCS, UBPC-; equity approach).	UNICA y Group facilitation/ catalyst group	Discussion groups, exchanges and debates.	-Level of participation -Collective protagonism -Dialogue of knowledge -Development of business plans. -Articulation of actors. -Conflict management. -Management skills. -Teamwork.	December 2022 - January 2023
Conciliatelocal demandswith policiesand strategies agricultural and livestock development policies and strategies.	Promote workshops for exchanges to reconcile the demands of the strategy with these identifications in the productive context.	Facilitation group, UNICA and GIAL	Discussion groups, exchanges and debates.	-Knowledge of the actors and the context and its structures. -Number of GIALs -Number of claims. -Number of farms. -Number of leaders. -Number of consultation meetings. -Number of interactive best practices workshops. -Number of discussion groups, exchanges and debates. -Number of events. -Quantity of audiovisuals, tutorials and support materials containing evidence of	
	Visit and exchange the experiences of advanced producers in local innovative processes or local enterprises.	Facilitation group, UNICA and GIAL	Productive context of action.		
	Generation of spaces for the socialization of accumulated experience and best practices.	Municipal government, Facilitation Facilitation Group, UNICA and GIAL	Visit to agro-ecological reference farms, exchange between producers, fairs, festivals, exhibitions, communication products, workshops, contests.		

	Participation in events organized by different mass and political organizations, ANAP, ATAF, MINAGRI, CITMA, FMC, UJC.	Municipal government, Facilitation Facilitation Group, UNICA and GIAL	Calls for grassroots, municipal, provincial, national and international events.	each multi-stakeholder action related to Food Sovereignty in the territory. - Number of publications of scientific articles in peer-reviewed, high-impact journals.	
	Exchange from Experiences with producers from other municipalities in Avila, the province and internationally.	Municipal government, Facilitation Facilitation Group, UNICA and GIAL	International diversity fair in Havana. Provincial and local fairs.	-Number of interviews with leading producers. -Number of festivals.	
	Dissemination of the operation, results, and impacts of the SIAL in the territory.	Mediaof communicationLocaland provincial media (Radio Morón, Invasor newspaper, Punto program on HEIs and ETECSA).	Communication products, television and radio spots.		
	Systematization of best practices.	Group facilitation group, UNICA, municipal government, GIAL	Briefing notes, executive reports, books, scientific articles, products, etc. communicative, tutorials.		

	Creation of audiovisuals, tutorials and support materials containing evidence of each multi-actor action.	Facilitation Group, UNICA, municipal government, GIAL,CUM	Audiovisuals and support materials.		
	Publication of scientific articles in peer-reviewed and high quality journals. impact.	CUM, UNICA, GIAL	References of good practices and materials for support.		
	Interview with leading producers who stand out for their good practices in Avila's TV and in Morón	GIAL, CUM, Municipal government.	Evidence of good practices of leading producers.		

Figure 15 below shows a scheme that summarizes the actions of the chronological steps proposed to strengthen the SIAL in the municipality of Morón.

Figure 15. Multi-stakeholder actions to strengthen the SIAL in the municipality of Morón.

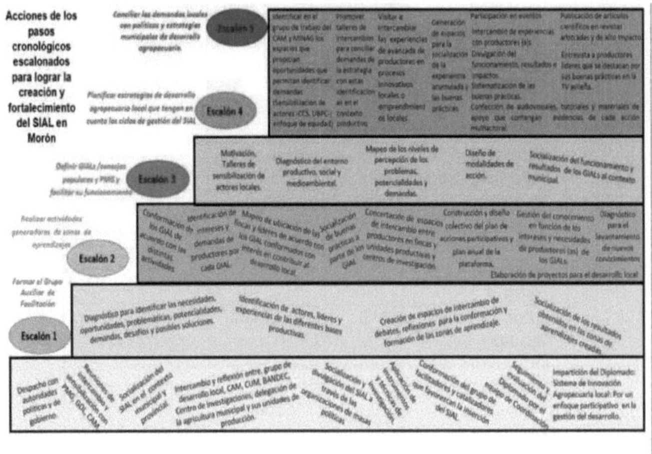

Source: Personal elaboration.

For innovation in a locality to last and be sustainable, it is necessary for the actors to be integrated into a coherent and solid platform and for there to be a close link and interaction between the municipal government-Science-academia and the productive sector to promote local development, but it must also be linked to the Integrated Community Action Work System (SITACI).All of the aforementioned imposes on the government of the municipality of Morón the will to act with a sense of commitment to face this challenge and advance in the rapid development of its main productive sectors, which will allow producers, actors and decision-makers to be active protagonists of the main transformations that will take place in their locality. Finally, Annex 4 presents the request made by the mayor of the municipality of Morón to the provincial coordination of the PIAL for the implementation of the SIAL in the territory.

CONCLUSIONS

1. The main characteristics of the agricultural context in the municipality of Morón in the province of Ciego de Avila are related to the fact that it has potentialities in the productive systems, but the inadequate integration among the actors causes a local development that does not evolve to higher stages.
2. The existence of systems of relations between local actors that are qualified from unilateral action and not in interconnectivity causes as main weakness innovation systems lacking the formation of knowledge networks and constant flow of information that favors Food Security.
3. A multi-stakeholder action plan was designed to be developed in the municipality of Morón by 2022, which responds to the implementation of the eight strategies of the SWOT matrices, based on the strengthening of the principles of the SIAL to contribute to Food Sovereignty, assuming the chronological steps to achieve the strengthening of the SIAL in the municipality of Morón.

RECOMMENDATIONS

The exploratory study conducted on the agricultural and innovation contexts of the municipality of Morón allowed us to suggest that it is necessary:

✓ Socialize among local stakeholders the requirements for the implementation of the multi-stakeholder action plan for strengthening the SIAL in the municipality of Morón.

✓ Determine indicators to measure the effectiveness of the strengthening of the SIAL in the municipality.

✓ The SIAL should be oriented to generate progress in food production, employment with inclusion, healthy and sustainable food, care of the environment, with adaptability to climate change, among other benefits through interactions, knowledge flows, learning, technology transfer, generating benefits in the agricultural sector and others to contribute to food security in the territory.

BIBLIOGRAPHIC REFERENCE

Aguilera, L. O., Rodríguez, A. M., Otero, Á. J., & Estupiñán, J. C. (2014). Projects, networks and sovereign functions in municipal university management of knowledge and innovation. Experiences from the province.

Alburqueque, 2003 Alburquerque, F. (2003). Course on local development. Course in Congress. Havana, Cuba.

Angelo, O (2002) Development issues for diversification. Digital format.

Arocena, R., & Sutz, J. (2006). The study of Innovation from the South and the perspectives of a New Development. Iberoamerican Journal of Science,
Technology, Society and Innovation (7).

Arocena, R., &Sutz, J. (2006). The study of Innovation from the South and the perspectives of a New Development. Iberoamerican Journal of Science, Technology, Society and Innovation (7).

Barreda. L, Bilbao. B, Martínez. Y, Castellanos. L, González. K. (2018). Final Work of the First Edition Diploma. Local Agricultural Innovation System: for a participatory approach in Development Management. Multi-stakeholder action plan to insert the Local Agricultural Innovation System in the Municipality of Morón.

Council of the administration of the municipality of Morón (2016). Comprehensive Development Project (PDI) of the municipality of Morón.

Coraggio, A (2002). A participatory approach to local development.

Digital format.

Diaz. Z, et al. (2017). Integrated sustainability evaluation system in Cuba. Case study: municipality of Morón, province of Ciego de Avila. CITMA Capacity Building Center in Morón.

Provincial Directorate of Physical Planning (2015). Esquema Provincial de Ordenamiento Territorial hasta el 2030 de la provincia Ciego de Ávila. Printed format.

Fis, Y., Arzola, L., and González, K. (2019). Local agricultural innovation system: a development alternative for the municipality of Baraguá based on a plan for the development of a local agricultural innovation system.

action from the conceptions of the culture of participation. Revista Universidad y Ciencia de la Universidad de Ciego de Avila, 8(2). Retrieved from http://revistas.unica.cu/index.php/uciencia/article/view/824/1922 (12/15/2020)
Food Security Forum, (2013) Challenges of the Future and Tasks of the Present Thursday 28 and Friday 29 November 2013 Food Security: Challenges of the Future Fernando Eguren CEPES. https://slideplayer.es/slide/1120285/ Accessed: September 2021
Forum Food Security: Challenges of the Future and Tasks of the Present Thursday, November 28 and Friday, November 29, 2013 Food Security: Challenges of the Future Fernando Eguren CEPES https://slideplayer.es/slide/1120285/ Accessed: September 2021
González, K (2017). Multi-stakeholder actions for the insertion of SIAL in the municipality of Venezuela. Mayabeque, Havana, 2017. Cuba
Jover JN, Ortiz HR, Proenza T, Rivas A. (2020) Higher education, science, technology and innovation policies and territorial development: new experiences, new approaches. Revista Iberoamericana de CTS; 15 (43): 187-218.
Marciniak, R (2012) How to do a SWOT Analysis? Business management Dr. Renata Marciniak's blog on strategies, models, management tools and other information needed to know how to manage a company https://renatamarciniak.wordpress.com/2012/10/04/analisis-dafo-example/ Accessed: September 2021
Núñez, J. (2007). Reflections on the social relevance of postgraduate education: How to build it? El nuevo conocimiento para la integración, Convenio Andrés Bello. Bogotá, No.3.
Núñez, J. (2014). University, knowledge, innovation and local development. Havana.
National Bureau of Statistics and Information (2012). Statistical Yearbook. Havana.

ONEI. Retrieved from: http://www.onei.cu. Accessed: September 2021.

National Bureau of Statistics and Information (2016). Statistical Yearbook. Havana. ONEI. Ciego de Avila, Morón.
National Bureau of Statistics and Information (2018). Statistical

Yearbook. Havana. ONEI. Ciego de Avila, Morón. 2019 edition.
Ortiz Pérez, H., Acosta Roca, R., Ruz Reyes, R., Arias, M., Rivas Diéguez, A., & Núñez Jover, J. (2021). Innovation system with a participatory approach in local development management. Sustainable pathway to increase food production, seeds and local welfare. Annals of the Academy From Sciences From Cuba,11(3),e1095.Retrieved from http://revistaccuba.sld.cu/index.php/revacc/article/view/1095/1239 Accessed: September 2021
Ortiz, R., la O, M., and Miranda, S. (2017). The Local Agricultural Innovation System. Conformation and functioning. In Romero, M. I., Caballero, R., Hernández, C. N., Núñez, J., Garcés, R., Ortiz, R., la O, R., Miranda, S., Roselló, T., Ríos, H., Cárdenas, R. M., Méndez, A., and Gil, Y. (Eds). Manual of the local agricultural innovation system. Towards a development management. Textos de apoyo al diplomado para la implementación del Sistema de Innovación Agropecuaria Local, 68-73. INCA Editions, National Institute of Agricultural Sciences, Local Agricultural Innovation Project (PIAL).
Ortiz, R; la O, Manuel and Miranda, Sandra: (2017). Curso Sistema de Innovación Agropecuario Local: conformación y formulación. Text to support the diploma course for the implementation of the Local Agricultural Innovation System. Mayabeque: Ediciones INCA. ISBN 978-959-7023-90-90-6ONEI 2015.
Ortiz. R, Acosta. R, Angarica. L and Guevara. F, (2017) CONTEXT DIAGNOSIS AND ATTITUDINAL CHANGE MONITORING FOR EFFECTIVE ACTIONS OF AN INNOVATION PROJECT. AGROPECUARIA. Cultivos Tropicales, 2017, vol. 38, no. 2, pp. 84-93 April-June, ISSN print: 0258-5936 ISSN digital: 1819-4087). Communist Party of Cuba (2011). Guidelines of the Economic and Social Policy of the Party and the Revolution. IV Congress of the Communist Party of Cuba.

CUBA'S FOOD SOVEREIGNTY AND NUTRITION EDUCATION PLAN (2020) REPUBLIC OF CUBA MINISTRY OF AGRICULTURE. The Havana, July 2020 Year 62 of the Revolution

Ramirez. B, Duharte. S, Mundiel. D, Jiménez. C and Arzola. L, (2019). Diploma Final Paper. Second Edition. provincial Local Agricultural Innovation System: for a participatory approach in Development

Management. Plan of actions for the creation of the GIAL in the productive bases of the Municipality of Morón, taking into account the production systems that integrate agriculture-livestock.
Rodríguez, A. (2020). The Local Agricultural Innovation System in Jobabo municipality (Cuba): actions for its strengthening. TERRA. Journal of Local Development, (7), 175-196. DOI 10.7203/terra.7.18660.
Tamayo. A, (2020) Agricultural production in Morón: from the furrow to the red area. Economics Infographics.02 May 2020 https://cubayeconomia.blogspot.com/2020/05/produccion-agricola-en-moron-del-surco.html Accessed: September 2021
Zamora.M, (2021)13 March2021Acknowledge vice-president Cuban Vice President agricultural potential of Ciego ofAvila. Newspaper Granma http://www.acn.cu/cuba/77698-reconoce-vicepresidente-cubano-potentialities-agricultural-potentialities-of-ciego-de-avila

ANNEXES

ANNEX 1 INTERVIEWS WITH WOMEN

Objective: To assess the position of women in Morón within the current agricultural context, in terms of their roles, access to products and services, self-improvement and decision-making.

1. Type of work performed by women and men and what role they play (reproductive, productive or community).
2. Positions held by women and men (explore the board of directors and roles within the cooperative).
3. If they receive stimulation, which one they receive and which one they receive.
4. In which decisions of the cooperative do women participate and in which do men participate?
5. Are women's practical needs met (bathrooms at production sites, clothes and shoes of appropriate size for women, work tools suitable for women, meeting schedules, etc.)?
6. Are training activities and events planned? Which ones involve women and which ones involve men? Who controls the main resources and inputs in the cooperative?

ANNEX 2
DISCUSSION GROUP (GOVERNMENT, CUM, FARMING COMPANY EXECUTIVES)

Objective: To identify the level of knowledge of local stakeholders in the agricultural context.

1. What are the main agricultural productions in the municipality?
2. Is there a municipal development strategy?
3. What is the place of agricultural development in the strategy?
4. What is the role of the municipal government?
5. What actions are proposed in the municipality to achieve agricultural development?
6. What are the main actors in agricultural production inthe municipality?
7. What other stakeholders should be incorporated and why?
8. Are the stakeholders articulated or do they work from their own objectives and agendas?
9. Is the agricultural development model proposed in the municipality sustainable? Why?
10. What type of agriculture do they implement (on agroecological basis?).
11. How would you rate the level of sustainability of the model and technologies implemented and why?
12. In your opinion, what are the main sustainability gaps (economic, social, technological and environmental) of the prevailing production system?

ANNEX 3
LOCAL STAKEHOLDERS DISCUSSION GROUP (GOVERNMENT, UNIVERSITY, PRODUCTIVE SECTOR WITH EMPHASIS ON CCS, UBPC, UEB, CPA)

Objective: To identify the level of knowledge of local stakeholders in the agricultural context.

1. Is there a space for consultation in the municipality?
2. Who participates, how often?
3. Are there agricultural strategies, programs and projects to promote local development? Identify the strategic lines.
4. Identify correspondence between local agricultural innovation priorities and innovative results.
5. Conduct a brief analysis of the main local productive arrangements (mini-industry, seed bank, biogas, biodigesters, local handicrafts, among others) present in the municipality.
6. Identify the actions promoted in the municipality towards Agricultural Development.
7. Identify the gaps that limit agricultural development.
8. Exemplify how the system that promotes agricultural development is working, how does participation, collective protagonism and dialogue of knowledge take place?
9. What are the exchange spaces that articulate with the Knowledge and Innovation Network for the development of the municipality of Morón?
10. How is it structured and how does it work? Mention the main results in the agricultural field.
11. What are the main sources of financing that exist in the municipality to promote local development?
12. What potentialities for knowledge management are identified in the municipality: teaching and research centers, other innovation actors in the municipality (even if they are not part of knowledge and innovation networks).
13. What relevant knowledge has been produced in the Morón municipalities.

14. What traditional knowledge exists in the municipality of Morón, how is it distributed and used?
15. Are there forms of training for local agricultural development actors such as this one in the municipality of Morón?
16. Do you consider relevant and necessary the implementation of forms of training aimed at local actors of agricultural development in your province? who is it for? how would it contribute to local agricultural development?
17. Who would you call on to motivate the idea and what are the steps you would have to take to achieve it?

ANNEX 4

Request from the mayor of the municipality of Morón to the provincial coordination of the PIAL for the implementation of the SIAL in the territory.

Fecha: 26-10-2021

A. La coordinación provincial del PIAL en Ciego de Ávila.

Por este medio se le demanda a la Coordinación del Proyecto Programa de Innovación Agropecuario Local (PIAL) la inserción del municipio en el mismo. Para la implementación del Sistema de innovación Agropecuario Local (SIAL) y se consolide las capacidades instaladas en función de la diversificación de la producción de alimentos y la gestión participativa del Desarrollo Local, a partir de la introducción de buenas prácticas surgidas del mismo en nuestro territorio.

Para eso, se debe institucionalizar procesos dentro de los gobiernos municipales, con la articulación de otros actores locales, agricultores e innovadores que accedan, reconozcan y utilizan las plataformas del SIAL para solucionar problemáticas de la producción agroalimentaria de su territorio.

Atentamente,

Nombre y Apellido del Intendente: Argelio Viche Gómez

Firma

Cuño

Printed by Books on Demand GmbH, Norderstedt / Germany